Human Factors Integration Challenges in the Terminal Radar Approach Control (TRACON) Environment

DOT/FAA/AR-02/127
DOT-VNTSC-FAA-02-11

Office of Aviation Research
Washington, DC 20591

Kim Cardosi, Ph.D.

U.S. Department of Transportation
Research and Special Programs Administration
John A. Volpe National Transportation Systems Center
Cambridge, MA 02142-1093

Final Report
January 2003

This document is available to the public
through the National Technical Information
Service, Springfield, VA 22161

U.S. Department of Transportation
Federal Aviation Administration

Notice

This document is disseminated under the sponsorship of the Department of Transportation in the interest of information exchange. The United States Government assumes no liability for its contents or use thereof.

Notice

The United States Government does not endorse products or manufacturers. Trade or manufacturers' names appear herein solely because they are considered essential to the objective of this report.

REPORT DOCUMENTATION PAGE		Form Approved OMB No. 0704-0188

Public reporting burden for this collection of information is estimated to average 1 hour per response, including the time for reviewing instructions, searching existing data sources, gathering and maintaining the data needed, and completing and reviewing the collection of information. Send comments regarding this burden estimate or any other aspect of this collection of information, including suggestions for reducing this burden, to Washington Headquarters Services, Directorate for Information Operations and Reports, 1215 Jefferson Davis Highway, Suite 1204, Arlington, VA 22202-4302, and to the Office of Management and Budget, Paperwork Reduction Project (0704-0188), Washington, DC 20503.

1. AGENCY USE ONLY (Leave blank)	2. REPORT DATE January 2003	3. REPORT TYPE AND DATES COVERED Final Report September 2001 – September 2002
4. TITLE AND SUBTITLE Human Factors Integration Challenges in the Terminal Radar Approach Control (TRACON) Environment		5. FUNDING NUMBERS FA3L1/A3112
6. AUTHOR(S) Kim Cardosi, Ph.D.		
7. PERFORMING ORGANIZATION NAME(S) AND ADDRESS(ES) U.S. Department of Transportation Research and Special Programs Administration John A. Volpe National Transportation Systems Center 55 Broadway Cambridge, MA 02142		8. PERFORMING ORGANIZATION REPORT NUMBER DOT-VNTSC-FAA-02-11
9. SPONSORING/MONITORING AGENCY NAME(S) AND ADDRESS(ES) U.S. Department of Transportation Federal Aviation Administration Office Aviation Research 800 Independence Ave., SW Washington, DC 20591		10. SPONSORING/MONITORING AGENCY REPORT NUMBER DOT/FAA/AR-02/127
11. SUPPLEMENTARY NOTES		
12a. DISTRIBUTION/AVAILABILITY STATEMENT This document is available to the public through the National Technical Information Service, Springfield, Virginia 22161.		12b. DISTRIBUTION CODE

13. ABSTRACT (Maximum 200 words)

This document describes human factors challenges that need to be considered in the implementation of planned enhancements to the Standard Terminal Automation Replacement System (STARS), Common Automated Radar Terminal System (ARTS), and the ARTS Color Display (ACD) in the Terminal Radar Approach Control (TRACON) environment. Some of the enhancements are tools that have been developed specifically to increase efficiency and capacity. Others provide information (regarding weather or aircraft position) that is more precise than the information currently available to controllers. The scope is limited to the air traffic control (ATC) specialist's workstation and specifically excludes Airways Facilities and Air Traffic Management issues. Issues are discussed within the TRACON environment and between environments, where applicable. The intent of this document is to pave the way for successful future integration efforts by identifying issues that need to be considered in the implementation process.

14. SUBJECT TERMS Human Factors, System Integration, TRACON, Decision Support Tools, Air Traffic Control, Automation, Passive Final Approach Spacing Tool, Controller-Pilot Data Link Communications			15. NUMBER OF PAGES 46
			16. PRICE CODE
17. SECURITY CLASSIFICATION OF REPORT Unclassified	18. SECURITY CLASSIFICATION OF THIS PAGE Unclassified	19. SECURITY CLASSIFICATION OF ABSTRACT Unclassified	20. LIMITATION OF ABSTRACT

Standard Form 298 (Rev. 2-89)
Prescribed by ANSI Std. 239-18
298-102

PREFACE

This research was sponsored by the Federal Aviation Administration's Office of the Chief Scientific and Technical Advisor for Human Factors (AAR-100). The author is extremely grateful to Paul Krois of that office for his critical review and substantial recommendations, and to Mike McAnulty and Kenneth Allendoerfer of the FAA Technical Center for their suggestions and corrections to an earlier draft. Finally, this study would not have been possible without the generous hospitality of several facilities, most notably the Atlanta, El Paso, Dallas-Ft. Worth, and New York TRACONs and to the Dallas-Ft. Worth Air Traffic Control Tower. We are very thankful to many controllers and managers who graciously shared their time, talents, experiences, and opinions with us.

METRIC/ENGLISH CONVERSION FACTORS

ENGLISH TO METRIC

LENGTH (APPROXIMATE)
- 1 inch (in) = 2.5 centimeters (cm)
- 1 foot (ft) = 30 centimeters (cm)
- 1 yard (yd) = 0.9 meter (m)
- 1 mile (mi) = 1.6 kilometers (km)

AREA (APPROXIMATE)
- 1 square inch (sq in, in^2) = 6.5 square centimeters (cm^2)
- 1 square foot (sq ft, ft^2) = 0.09 square meter (m^2)
- 1 square yard (sq yd, yd^2) = 0.8 square meter (m^2)
- 1 square mile (sq mi, mi^2) = 2.6 square kilometers (km^2)
- 1 acre = 0.4 hectare (he) = 4,000 square meters (m^2)

MASS - WEIGHT (APPROXIMATE)
- 1 ounce (oz) = 28 grams (gm)
- 1 pound (lb) = 0.45 kilogram (kg)
- 1 short ton = 2,000 pounds (lb) = 0.9 tonne (t)

VOLUME (APPROXIMATE)
- 1 teaspoon (tsp) = 5 milliliters (ml)
- 1 tablespoon (tbsp) = 15 milliliters (ml)
- 1 fluid ounce (fl oz) = 30 milliliters (ml)
- 1 cup (c) = 0.24 liter (l)
- 1 pint (pt) = 0.47 liter (l)
- 1 quart (qt) = 0.96 liter (l)
- 1 gallon (gal) = 3.8 liters (l)
- 1 cubic foot (cu ft, ft^3) = 0.03 cubic meter (m^3)
- 1 cubic yard (cu yd, yd^3) = 0.76 cubic meter (m^3)

TEMPERATURE (EXACT)
$[(x-32)(5/9)]$ °F = y °C

METRIC TO ENGLISH

LENGTH (APPROXIMATE)
- 1 millimeter (mm) = 0.04 inch (in)
- 1 centimeter (cm) = 0.4 inch (in)
- 1 meter (m) = 3.3 feet (ft)
- 1 meter (m) = 1.1 yards (yd)
- 1 kilometer (km) = 0.6 mile (mi)

AREA (APPROXIMATE)
- 1 square centimeter (cm^2) = 0.16 square inch (sq in, in^2)
- 1 square meter (m^2) = 1.2 square yards (sq yd, yd^2)
- 1 square kilometer (km^2) = 0.4 square mile (sq mi, mi^2)
- 10,000 square meters (m^2) = 1 hectare (ha) = 2.5 acres

MASS - WEIGHT (APPROXIMATE)
- 1 gram (gm) = 0.036 ounce (oz)
- 1 kilogram (kg) = 2.2 pounds (lb)
- 1 tonne (t) = 1,000 kilograms (kg) = 1.1 short tons

VOLUME (APPROXIMATE)
- 1 milliliter (ml) = 0.03 fluid ounce (fl oz)
- 1 liter (l) = 2.1 pints (pt)
- 1 liter (l) = 1.06 quarts (qt)
- 1 liter (l) = 0.26 gallon (gal)
- 1 cubic meter (m^3) = 36 cubic feet (cu ft, ft^3)
- 1 cubic meter (m^3) = 1.3 cubic yards (cu yd, yd^3)

TEMPERATURE (EXACT)
$[(9/5) y + 32]$ °C = x °F

QUICK INCH - CENTIMETER LENGTH CONVERSION

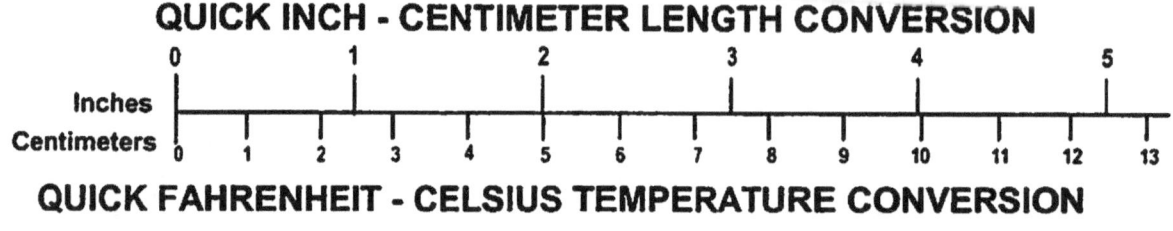

QUICK FAHRENHEIT - CELSIUS TEMPERATURE CONVERSION

°F	-40°	-22°	-4°	14°	32°	50°	68°	86°	104°	122°	140°	158°	176°	194°	212°
°C	-40°	-30°	-20°	-10°	0°	10°	20°	30°	40°	50°	60°	70°	80°	90°	100°

For more exact and or other conversion factors, see NIST Miscellaneous Publication 286, Units of Weights and Measures. Price $2.50 SD Catalog No. C13 10286

TABLE OF CONTENTS

Section	Page
1. INTRODUCTION AND SCOPE	1
2. TRADITIONAL HUMAN-SYSTEM INTERACTION (HSI) INTEGRATION ISSUES	3
3. GENERAL DESCRIPTION OF TRACON INTEGRATION ISSUES	7
4. SPECIFIC TRACON ENHANCEMENTS AND ASSOCIATED INTEGRATION CHALLENGES	9
4.1 CTAS	9
4.1.1 System Description	9
4.1.2 Traditional Human Factors Integration Issues Associated with pFAST	10
4.1.3 Other Integration Issues Associated with pFAST	11
4.1.4 pFAST Integration Issues Across Environments	13
4.2 Controller Automation Spacing Aid (CASA)	14
4.2.1 System Description	14
4.2.2 Integration Issues to be Considered	14
4.3 Integrated Terminal Weather System (ITWS)	15
4.3.1 System Description	15
4.3.2 Integration Issues to be Considered	15
4.4 Controller-Pilot Data Link Communications (CPDLC) and Next Generation Digital Air-Ground Voice Communications (NEXCOM)	16
4.4.1 System Description	16
4.4.2 Integration Issues to be Considered	16
4.5 Enhanced Surveillance	19
4.5.1 System Description	19
4.5.2 Integration Issues to be Considered	19
4.6 Enhanced Traffic Management System (ETMS) Upgrade	20
4.6.1 System Description and Integration Issues to be Considered	20
5. THE NEW FACE OF HUMAN-SYSTEM INTEGRATION (HSI)	25
6. SUMMARY OF EXISITNG CHALLENGES AND SUGGESTIONS FOR FURTHER RESEARCH	31
APPENDIX	35
GLOSSARY	37
REFERENCES	39

This page intentionally left blank

1. INTRODUCTION AND SCOPE

This document describes human factors issues that need to be considered in the implementation of planned enhancements in the Terminal Radar Approach Control (TRACON) environment. The scope is limited to the air traffic control (ATC) specialist's workstation and specifically excludes airways facilities and air traffic management issues. The components comprising the legacy systems in the TRACONs are the Standard Terminal Automation Replacement System (STARS), Common Automated Radar Terminal System (ARTS), and the ARTS Color Display (ACD). The planned enhancements discussed in this document were primarily derived from the description of pre-planned product improvements (P^3Is) presented in the Federal Aviation Administration's 1999 National Airspace System (NAS) Architecture. Issues are discussed within the TRACON environment and between environments, where applicable. This forecast of integration issues assumes that STARS, Common ARTS, ACD, and all enhancements (subsystems) are fully functional and perform "as advertised." This document is not intended to be an evaluation of STARS, the ACD, or any of the enhancements discussed; nor does it detract from the necessity to evaluate how the enhancements should be implemented onto the specific legacy systems – STARS (Full Service Level or Early Display Capability) or Common ARTS and ACD. The sole intent of this document is to pave the way for successful future integration efforts by identifying issues that need to be considered in the implementation process.

Descriptions of systems were derived from published reports and government documents. Operational experiences were determined through published reports and interviews with controllers, managers, and training personnel at Atlanta, Dallas-Ft. Worth and El Paso TRACONs.

This page intentionally left blank

2. TRADITIONAL HUMAN-SYSTEM INTERACTION (HSI) INTEGRATION ISSUES

System integration in the NAS has many facets. Subsystems, such as decision support tools, need to be accepted by the workforce and properly integrated into the controller's workstation to ensure that they are used effectively. Traditionally, work on system integration has focused on the human-system interface (HSI) for the new and legacy systems and was limited to within a workstation or any set of equipment that any one person would be expected to use. It is these traditional issues that will be explored first.

Within ATC environment, there are integration issues that can be considered global or "top-down." These issues affect the controller at the facility or "work culture" level, rather than at the level of the individual workstation. Globally, integration issues have implications for staffing levels, training requirements, controller roles and responsibilities and teamwork. Most of the traditional integration issues, however, can be described as local or "bottom-up." At the lowest level, the integration of subsystems into the legacy system directly affects the controllers' tasks and the design of the controller workstation. How well the subsystems are integrated can affect controller efficiency, workload, and job satisfaction issues, as much as the performance of the individual subsystems. These bottom-up issues deal mostly with helping to ensure that the controller's information requirements are met for their tasks. They include all aspects of information presentation and organization, and controls for manipulating information (such as function keys). In general, the traditional human factors integration issues that are examined within a workstation include the following:

- Is the information from the subsystem displayed appropriately so that it displays only the necessary information as it is needed, and neither obscures critical information or is unduly distracting?

- Is the information displayed compatible with other information displayed, or available, to the controller?

- Are the procedures (e.g., use of function keys) to be used with the subsystem compatible with the procedures required by other systems the controller uses?

- Are critical failure modes adequately displayed to the controller and propagated through the relative systems?

Each of these questions will now be discussed.

Is the information from the subsystem displayed appropriately so that it displays only the necessary information as it is needed, and neither obscures critical information or is unduly distracting?

Whenever information is added to complex displays, such as the controllers' situation display, it usually adds to the degree of display "clutter" and has the potential risk of obscuring, or otherwise detracting from, important information. In this sense, there are potential perceptual and attentional "costs" to any new information presented. If the new information adds clutter to the extent that other information is difficult to read (such as with overlapping data blocks), this is a perceptual price; if the new information detracts the users attention from other important information, then there is an attentional price. Any new information presented on the situation display must be worth the perceptual and attentional "price" of displaying it. This means that the controller should be presented with, and *only* with, information that is useful in performing the required tasks at the time it is required. Some capabilities (such as with free form text) will permit the controller to input and place alphanumeric text on the STARS TCW display. While such a tool may be useful to replace the handwritten notes used during a controller position relief briefing, controllers will need to be given recommendations and cautions for use, to avoid problems of display clutter and overlapping text and symbols. Effects on the efficacy of the position relief briefing should also be explored to determine whether the capability is worth the attentional price.

If the information is such that immediate action is required, or the information is required to make tactical decisions, (such as aircraft position), then the information needs to be displayed continuously. Obscuration of critical information was an issue with the initial STARS prototype. The original STARS display used opaque windows to convey (even the most mundane) information. These windows blocked portions of the situation display and remained until the controller responded to them. Consequently, the window could have observed a conflict alert or other critical inforamtion. With the implementation of any system or subsystem, care must be taken so that critical information is not obscured.

If the information is to be used to make strategic decisions (where no immediate action is required), then the information should be available to the controller, but not continuously displayed on the situation display. This issue is most critical when it involves Decision Support Tools (DSTs).

Is the information displayed compatible with other information displayed, or available, to the controller?

As capabilities to present new information to controllers evolve, care must be taken to ensure that the new information is at least as good as the old information and either replaces the other information presented from other sources or is compatible with it. Examples of this include information from different sources regarding weather, aircraft position, and potential conflicts. In some cases, the controller may be able to select the source of the information. If so, the source of the information must be clear to the controller.

Are the procedures used for data entry and recall (e.g., use of function keys) to be used with the subsystem compatible with the procedures required by other systems the controller uses?

It is sometimes the case that when new subsystems are implemented, controllers lose some of the functionality (e.g., "slew and enter" capabilities) that they have become accustomed to. In considering the HSI of the integration of a new capability, it is important to determine that the same functions (e.g., range selection) are selected in the same way, and that any functionality that is lost is compensated for in an acceptable usable way.

Are critical failure modes adequately displayed to the controller and propagated through the relative systems?

Subtle failures such as a display's temporary inability to update need to be displayed to the controller. Also the ways in which the failure affects other systems need to be conveyed to the controller. For example, if a radar is temporarily out and no back-up is available, the controller needs to be informed that the radar is out and any decision support tools that rely on radar input should echo this limitation.

This page intentionally left blank

3. GENERAL DESCRIPTION OF TRACON INTEGRATION ISSUES

The growth of air traffic, combined with the increased demand for flexibility and elimination of current restrictions, requires enhancements in the terminal architecture to enable the controller to maintain safety and increase efficiency. The successful integration of any subsystem into the legacy system is necessary for the full realization of the expected benefits projected to be afforded by the subsystem. All system benefits are projected on assumptions of a given level of human performance. This includes correct and efficient data entry, correct interpretation of displayed data, correct and efficient response input, etc. Systems that are poorly designed or poorly integrated can induce user errors. Such errors can lead to poor user acceptance, poor system performance and unrealized system benefits. As subsystems are developed to address specific operational needs, their development is usually independent of the evolution of the legacy system. This means that the issues surrounding the integration of the systems are usually not identified until the first stages of operational evaluation.

While general human factors integration issues regarding the implementation of subsystems would be expected to be similar across STARS and ACD, each implementation will need to be the examined separately in the specific context within which they will be implemented. Facilities vary widely on many dimensions, ranging from the characteristics of the traffic mix to ambient light levels at the controller's workstation.[1] Consequently, close attention needs to be paid to integration issues within each facility.

There are also integration issues across environments. Information (e.g., regarding aircraft position) and conflict resolution advice provided to en route controllers must be compatible with that provided to the TRACON controllers with whom they coordinate and interact. Similarly, information provided to tower controllers must be compatible with that provided to the TRACON controllers. Finally, for maximally efficient operations, the information provided to controllers in the oceanic sectors should be of the same quality as that provided to en route controllers.

In addition to the integration issues surrounding systems within and between ATC environments, there are also air-ground issues to be considered. Cockpit systems that command a pilot to maneuver will have implications for TRACON controllers. If the information regarding the position of potential threat aircraft that is provided to the pilot is substantially different from the information provided to the controller, errors and inefficiencies can result. Any cockpit system (such as the Traffic Alert and Collision Avoidance System – TCAS) that will result in a pilot maneuvering without a specific ATC instruction to do so will also have implications for ATC. This includes cockpit systems that support shared-separation responsibility, so that an aircraft may maneuver without a specific ATC instruction, and collision-avoidance systems that may require the pilot to maneuver the aircraft before informing ATC of the altitude or heading change. Finally, ATC systems that seek to enhance efficiency will also have implications for the flight deck. The timing of presentation of information such as runway assignments to the cockpit

---

[1] Measures of ambient light levels vary widely from facility to facility. Average measurements taken at the controller workstation varied from 11.15 fc (at El Paso), 2.7 fc at the Atlanta TRACON, and .09 fc at the Dallas-Ft. Worth TRACON. See the Appendix for a complete description of these measurements.

has a dramatic effect on pilot workload and may even affect the probability of a runway incursion as it affects the time available to anticipate taxi routes and runway crossings.

These integration issues within and across environments will now be discussed in more detail in the context of specific TRACON enhancements in Section 4.

4. SPECIFIC TRACON ENHANCEMENTS AND ASSOCIATED INTEGRATION CHALLENGES

Significant improvements in basic capabilities have been initiated in the last several years. These include: improved radar processing capabilities (as provided by Common ARTS and STARS), display replacement (STARS and ACD), and refinement of the conflict alert (CA) and Minimum Safe Altitude Warning (MSAW) algorithms (Beal, Reid, and Schlimper, 2000). Many other enhancements are planned to accommodate the changing needs of the air traffic community. These enhancements are at various stages of maturity and include: improved weather information, improved communication equipment, and new DSTs. In the TRACON environment, the most significant subsystems proposed are:

- Center TRACON Automation System
  - passive Final Approach Spacing Tool (pFAST)
  - active Final Approach Spacing Tool (aFAST)

- Controller Automation Spacing Aid (CASA)

- Integrated Terminal Weather System (ITWS)

- Controller-Pilot Data Link Communications (CPDLC) and Next Generation Air-Ground Communication System (NEXCOM)

- Enhanced Traffic Management System (ETMS) Upgrade

Some of these enhancements are tools that have been developed specifically to increase efficiency and capacity. Others provide information (regarding weather or aircraft position) that is more precise than the information currently available to controllers. Each proposed subsystem projects benefits based on the optimal use of this subsystem. However, these benefits can only be realized in actual operations if the subsystem is properly integrated into the controller workstation. Each of these will be examined with respect to potential integration issues.

4.1 CTAS

4.1.1 System Description

Center TRACON Automation System (CTAS) is a suite of tools designed for use by TRACON and Air Route Traffic Control Center (ARTCC) controllers. The CTAS tools designed for the TRACON are passive Final Approach Spacing Tool and active Final Approach Spacing Tool. PFAST presents the controller with a runway assignment and sequence number in an additional line to the data block; aFAST would supplement this with heading and speed recommendations. PFAST has had limited implementation that has resulted in a wealth of operational experience. However, the future implementation plan for pFAST and aFAST is unclear. The Free Flight Phase 2 Research Program Plan (November 2001) identifies another DST for the TRACON known as Expedite Departure Path (EDP). This tool is described as under development by NASA to provide speed, heading and altitude advisories to controllers to help balance the traffic over departure fixes, allow for expedited climbs and more efficient routing into the en-route stream. As the departure counterpart to aFAST, such a tool would be expected to have the same human factors and operational issues as pFAST and aFAST.

Initially, there was strong praise for pFAST and its development process from both the National Research Council (Wickens, Mavor, Parasuraman, and McGee, 1998) and the National Air Traffic Controllers Association (NATCA). Controllers and human factors specialists working closely with the engineers as a team as they continued to refine product was regarded as a highly successful development and deployment strategy. The initial operational experience was also regarded as successful. In one study, use of pFAST resulted in an increase in acceptance rates of 2.5 aircraft per hour at Dallas-Ft. Worth TRACON (Meyer, Post, Blucher, and Fralik, 2000). The median peak throughput increased from 105 to 109 operations per 30-minute period under instrument approaches and from 111 to 114 operations when using visual approaches (ibid). Just as important as the increase in capacity is the fact that such increases in throughput were achieved within acceptable limits for workload (Lee and Sanford, 1998).

In addition to the demonstration of increased capacity, this initial experience also showed that many controllers had confidence in the system as defined by the advisories being judged as acceptable by the controllers (Davis, Isaacson, Robinson, den Braven, Lee, and Sanford, 1997). This is an important accomplishment in the development of a system, since controller confidence is critical to ensure that the system is used as intended and projected benefits are realized. The general consensus was that the system "thought" like a controller and provided advisories that the controllers agreed with. In fact, Dick Swauger, the national technology coordinator for NATCA at the time said that using the tool was "like having a top controller at your side whispering in your ear… it does make good controllers better" (Perry, 1997, p. 31). Now, however, the system is not being used. The reasons for this are varied and complex and will now be explored.

The initial implementation experience of pFAST was an extremely rich learning experience and was successful in many respects. There has been much discussion regarding the lessons to be learned from this experience. Such lessons can help to identify integration issues that should be considered in the development and implementation of future decision support tools.

4.1.2 Traditional Human Factors Integration Issues Associated with pFAST

The initial operational experience of controllers using pFAST at the Dallas-Ft. Worth TRACON with three primary arrival runways was regarded as highly successful from both an engineering and human factors perspective. After the field evaluation was complete, the more complex Metroplex configuration with its fourth arrival runway was implemented. The additional runway added to the complexity of the traffic flows, resulting in more data blocks being displayed in a given space on the screen. The use of pFAST increased the amount of information displayed in the data block as it added another line to the (2-line) data block. Use of pFAST also increased the complexity of the air traffic flows; this is necessary to increase efficiency. Thus, the use of the additional runway resulted in more data blocks being displayed in the immediate vicinity of one's own data blocks, and the use of pFAST resulted in larger data blocks. These two factors resulted in a noticeably larger proportion of overlapping and obscured data blocks.

This part of the pFAST experience points to two mainstream human factors issues associated with the integration of new subsystems on the legibility of data blocks. First, subsystems that increase the efficiency of operations by filling gaps between aircraft, increase the complexity of operations (as timing of individual maneuvers becomes more critical) and decrease the space between data blocks; this adds to the display clutter. Increased complexity makes it more difficult to search for a particular aircraft under one controller's jurisdiction when they are necessarily displayed among many others that can only be differentiated by a position symbol and aircraft ID. Increased traffic complexity is also a well-known cause of operational errors. Second, the increase in the number of aircraft in the immediate vicinity also causes more data blocks to overlap and become partially or totally obscured. There are many information display techniques that could be used to help alleviate this problem. For example, it is clear that controllers need a way of differentiating their own data blocks from others' that is more effective than the current cue (i.e., the position symbol). One way to do this would be to color code the position symbol (or another portion of the data block). Color coding the entire data block (as is currently employed in ACD) may not be recommended because there is anecdotal evidence to suggest that this could increase the probability that a controller will fail to detect a conflict between an aircraft under their control, i.e., in "their color," and an aircraft in another color (see Cardosi and Hannon, 1999). Second, controllers need to be able to make an adjustment to their data blocks to make them legible when they are superimposed on another controllers' data block. One intuitive solution to this is to give controllers the ability to differentially control the brightness (intensity) of their data blocks. Controllers using the ACD currently have this capability. However, once again, controllers need to be cautioned that this display technique should only be used as needed to increase the legibility of the data blocks; displaying one's own traffic at a higher intensity than the other aircraft on the display increases the probability that the lower intensity aircraft might unintentionally be ignored. The degree of difference in intensity (brightness) that could affect the controller's ability to detect potential conflicts needs to be empirically determined. Research is needed to evaluate different techniques to help the controller identify "their" data blocks without increasing the probability that the controller would fail to detect a potential conflict between an aircraft under their control and an aircraft not under their control.

4.1.3 Other Integration Issues Associated with pFAST

In addition to the traditional types of HSI integration issues that surfaced as a result of the facilities' experience with pFAST, other critical integration issues also emerged. It is well known that the level of controller confidence in a tool is pivotal in the tool's success or failure. This confidence is largely determined by how accurate and reliable the tool is perceived to be. This is, in turn, largely dependent on how well the system is adapted to the specific site and its operations. Site adaptation is almost always more extensive than originally anticipated, but is *the* determining factor in the success or failure of a decision support tool, and cannot be short-changed due to scheduling or other artificial constraints.

The operational experience with pFAST was complex and changed over time. With full implementation, the additional runway in use, and more controllers using the system, controller confidence in pFAST was varied. The degree of confidence that an individual controller had in pFAST seemed to be a function of how well pFAST worked in that particular sector/area. Another key point is that pFAST was originally conceived, and presented to controllers, as an advisory system (which *usually* means that it can be used or not as the controller sees fit). However, the system works best when all controllers use the runway assignments offered; in fact, system performance is severely degraded when a large number of the advisories presented are not used. Requiring controllers to use the advisories is only acceptable as long as the advisories are ones that the controllers would be inclined to use, that is, ones that the controllers consider to be good recommendations (or at least workable solutions that do not substantially increase workload). If, however, the controllers think that the advisories are problematic and are "forced" to use them anyway, then a primary rule of the proper allocation of function between the user and the automation is violated. No automated system can be expected to have the wealth and breadth of information that the controller does. Nor can any DST be expected to be able to exhibit the same degree of flexibility in decision making amid rapidly changing information that the controller does routinely. The controller must remain in the position of being the ultimate decision-maker and should never be a slave to the automation.

In order to be able to evaluate the pFAST advisories, it is necessary to observe the system with all of the advisories implemented. To this end, the Dallas-Ft.Worth TRACON instituted a trial period during which controllers were instructed not to change a (pFAST suggested) runway assignment without supervisory approval. This did not help to endear pFAST to the controllers that worked sectors in which pFAST runway assignments problematic (for reasons specific to the sector operations, such as satellite operations). In hindsight, it would have been better to continue to refine the site adaptation (to help ensure the operability in all sectors) before implementing it in the new, more complex runway configuration.

The intent of pFAST is to increase capacity by helping controllers to increase the efficiency of their operations. With more attention to site adaptation and a continuous cadre of operational, human factors and engineering support, pFAST has potential for being re-engineered as a viable DST that could enhance efficiency and safety. This will be a necessary prerequisite for any progress toward the implementation of active FAST. aFAST proposes to enhance this capability by providing the controller with recommended speed and heading adjustments. The most important issue to be resolved with aFAST is operational confidence in these advisories. The strong consensus amongst controllers was that there was no confidence in the sequence numbers assigned by pFAST. If the system could not be "trusted" to give reliable sequence numbers, it seems unlikely that it would offer usable speed and heading adjustments. If aFAST is to be viable, it will need to be demonstrated to be highly accurate and reliable under all operational conditions (e.g., in different wind and weather conditions, with aircraft in holding patterns, increases in spacing requirements, satellite operations) before initial implementation. This will require iterative stages of interactive testing with controllers, system developers, and human factors specialists.

If aFAST proves viable, the issues associated with how the information is to be displayed to the controller must be explored. For example, a recommended heading adjustment could be displayed as a change (e.g., turn 30 degrees to the right) or as new heading (heading 230). Where the information is displayed on the controllers screen is another integration issue. One possibility is that the speed and heading would time-share with the current third line on the data block; this would make the original pFAST information of runway assignment and sequence available only half the time. Another possibility is that it would occupy a fourth line on the data block; this would further add to display clutter. Display options will need to be systematically evaluated so as to successfully integrate aFAST into the controller's workstation.

4.1.4 pFAST Integration Issues Across Environments

If pFAST could be improved to function "as advertised" under all operational conditions, use of pFAST would be of great benefit to pilots. Knowing the arrival runway early in the approach allows pilots to perform the necessary programming of cockpit systems and conduct the pre-arrival planning (e.g., approach briefing) early in the arrival process. Thinking about the possible taxi routes from the arrival runway and the potential for incursions (e.g., determining whether there is a runway between the arrival runway and the gate) is a critical step in helping to prevent runway incursions. Conducting these activities as early as possible, preferably prior to descent, is an effective workload management strategy that allows for the pilots' full attention to be focused on the tasks of stabilizing the approach, landing, rolling out, and taxiing off the runway. Conversely, last minute runway changes can be very disruptive and increase pilot workload dramatically. In addition to the natural disruption of a "last minute" change of plans, the aircraft's flight management system may require reprogramming to capture the localizer or selected approach to the newly assigned runway (particularly if the runway assignment has changed more than once). Having the runway assignment prior to descent and not subject to change (as is often the case at some airports) could be a substantial benefit to pilots and could help reduce the number of runway incursions.

There are other integration issues associated with pFAST that go beyond the boundaries of the TRACON environment. One of these is the effect of the use of pFAST on tower operations. Informal interviews with DFW tower controllers revealed that the period of "mandated" pFAST use was not only noticed, but welcomed. Use of pFAST resulted in more evenly distributed arrival flows and a perception that capacity could be increased without a concomitant increase in workload.

A more subtle integration issue that crosses operational environments is the interoperability of CTAS and User Request Evaluation Tools (URET). The algorithms for aircraft trajectory modeling used by the terminal functions of CTAS and the en route URET are fundamentally different (Ryan, Kazunas, Paglione, and Cale, 1997). While it is not necessary for the two tools to provide the same "advice," it would not be acceptable for the tools to provide radically different information about the same aircraft to two different users (e.g., a TRACON controller coordinating with an en route controller). Also, the use of one tool in one environment (TRACON or ARTCC) should not adversely affect either the use of the other tool in the other environment.

4.2 CONTROLLER AUTOMATION SPACING AID (CASA)

4.2.1 System Description

Converging Runway Display Aid (CRDA) is the first implementation of Controller Automation Spacing Aid) CASA. Like CTAS, such tools are also designed to increase capacity by making operations more efficient. By helping controllers to visualize the spacing between aircraft on converging approaches in terminal airspace, spacing between aircraft can be strategically reduced. CRDA helps controllers sequence traffic for arrival on converging runways during instrument meteorological conditions. CRDA shows the aircraft on the approach paths to both runways - the aircraft on approach and a "ghost" image of the aircraft on approach to the other runway. This image helps the controller to judge the separation more effectively so that tighter spacing can be maintained even in periods of low visibility. Note that the intent of this system is not to provide decisions to controllers, but rather to provide information to controllers that enhance their ability to make decisions regarding spacing.

CRDA is similar in function to the Precisionl Runway Monitor (PRM), a dedicated display and control position designed to enable closely spaced parallel approaches in poor visibility. This position resides in the TRACON, but can override the tower frequencies, if necessary. At this position, the controller issues correction instructions to keep pilots out of the "No Transgression Zone" (NTZ) between the two runways. If a pilot does enter the NTZ, the aircraft on the parallel approach (i.e., the non-transgressor) is sent around to avoid a conflict. Since PRM is a dedicated position, there are no integration issues for the TRACON per se. However, since the impact of the position resides in the tower, integration issues with the tower should be revisited. For example, should the position physically reside in the tower? Is there a way to have the minor adjustment control instructions issued that may be less intrusive to tower operations than overriding the frequency?

4.2.2 Integration Issues to be Considered

Since CRDA is a sequencing tool, it will need to be compatible with pFAST or similar DSTs used in the TRACON environment. CRDA helps controllers judge the spacing between aircraft on converging runways more effectively than is otherwise possible in periods of low visibility. It is still incumbent upon the controller to adjust the spacing between aircraft. This task becomes more challenging as the descent speeds of the two aircraft diverge. If CRDA evolves from a visualization tool to a more proactive DST, it may be useful to consider incorporating an additional algorithm, similar to the Descent Advisor (DA) designed for the en route environment, which would take the aircraft descent speeds into account as it provides guidance on maintaining the desired spacing.

4.3 INTEGRATED TERMINAL WEATHER SYSTEM (ITWS)

4.3.1 System Description

The Integrated Terminal Weather System will provide higher quality weather information (in graphic and text format) to TRACON controllers than is currently available. Significant human factors effort has been expended to determine the best way to present this information on STARS. (Allendoerfer, Bacon, Bohne, and Freitag, 2001). A multi-disciplinary working group, made up of representatives from user groups (the National Air Traffic Controllers Association and the Professional Airway System Specialists [PASS]), FAA operations, requirements, acquisition, and human factors specialists was conveined. The working group developed a prototype presentation format for ITWS that combined the color and format of the presentation of traffic in STARS with the ITWS presentation format. Presentation of ITWS on the ACD will need similar consideration.

Currently, STARS and ACD use different combinations of colors and levels of density of texture patterns (known as "stipple patterns") to create six levels of precipitation. STARS uses two colors (dark gray-blue and mustard) and three levels of stipple (none, sparse and dense); ACD uses three colors (gray, orange and red) and two levels of stipple. (ITWS will present precipitation information in these formats. However, ITWS is capable of presenting a considerably higher level of detail of weather information that includes storm cells, microbursts, and gusts fronts. ITWS also presents several alerts (such as for microbursts). This detailed information would require the additional presentation of more combination of hues and intensities.

4.3.2 Integration Issues to be Considered

Weather information is important to controllers; however, the detail of weather information presented must be determined by operational requirements. Any information presented on the situation display must be directly operationally relevant and immediately useful. While the source of the information must be as accurate as possible, it is counterproductive to present more detailed information than is necessary. The weather information that is capable of being presented on ITWS is much more complex than what is currently displayed. Such complex information has a cost in terms of potential distraction or obscuration of more important information. While tremendous effort has gone into determining the best way to present ITWS information on STARS, nothing could be found to document a similar effort to determine the operational requirements for all of the information that ITWS provides. If no formal task analysis has been conducted to determine the specifics of the type and detail of weather information that TRACON controllers need, and how this information would be used to make specific decisions, it cannot yet be determined whether the ITWS information will be worth the perceptual and attentional costs of displaying it. It may be the case that such detailed information is best presented to Traffic Management Units (TMUs) and to shared ETMS displays.

Another planned enhancement to terminal weather is the "Integrated Turbulence Forecast Algorithm" (IFTA). Currently, controllers depend on pilot reports for identifying areas of turbulence. IFTA will display "areas of turbulence with different colors which represent different forecast intensities" (Jones, 2001, p.125). Again, while this information is potentially useful to controllers, it is not clear that it is information that should be continuously displayed on the situation display.

4.4 CONTROLLER-PILOT DATA LINK COMMUNICATIONS (CPDLC) AND NEXT GENERATION DIGITAL AIR-GROUND VOICE COMMUNICATIONS (NEXCOM)

The TRACON environment presents serious challenges in handling the anticipated volume of voice communications. In a 1996 study of TRACON voice communications, there was an average of 4.5 controller transmissions per minute per frequency, containing an average of 3.3 clearances per minute (Cardosi, Brett, and Han, 1996). This can be contrasted with a similar study of en route communications that showed an average of 1.8 controller transmissions per minute, containing an average of 1.3 clearances per minute (Cardosi, 1993). Because of this high volume, there is a relatively high number of communication errors per hour, even though the overall communication rate is very low. With one percent of the controllers' instructions resulting in a readback error, and 60 percent of these readback errors corrected, there was an average of one uncorrected readback error every 1½ hours on TRACON frequencies (Cardosi, Brett, and Han, 1996). By contrast, the average in the en route environment was one uncorrected readback error every 13 hours (Cardosi, 1993). Frequency congestion and associated problems (such as blocked or "stepped on" transmissions) in the TRACON environment could threaten the realization of expected benefits from other subsystems that are dependent upon effective voice communications.

4.4.1 System Description

Controller-Pilot Data Link Communications provides a sorely needed alternative to voice communication between pilots and controllers. It affords the capability to "uplink" information (instructions, frequency changes [transfer of communication], etc.) to the cockpit and "downlink" information (requests, acknowledgements, etc.) to the controller. It also allows some routine transmissions, namely frequency changes, to be automated with the handoff. CPDLC is scheduled to be implemented in the TRACON environment after full implementation in the en route environment. NEXCOM purports to increase the efficiency and capacity of air-ground voice communication, in part by providing digital modulation. The first implementation of NEXCOM is expected to use the VHF Digital Link (VDL) Mode 3 protocol and has the potential for multiplying the number of available voice channels up to four (Kabaservice, 1998). NEXCOM will operate in parallel with the present analog voice system and is expected to be implemented in high and ultra-high altitude sectors by 2008. Selected high density terminal sectors are scheduled to transition to digital NEXCOM by 2015. (NAS Architecture, 1999.)

4.4.2 Integration Issues to be Considered

The first implementation of NEXCOM will use VDL Mode 3 that integrates voice and data. Aircraft and ATC facilities in Europe and other parts of the world are planning to implement

VDL Mode 2 (data only). Since CPDLC will be used to downlink information, such as aircraft position, route information, etc., to controllers, as well as uplink information to the cockpit, integration issues associated with aircraft downlinking information with Mode 2 and trying to receive information transmitted via Mode 3 should be anticipated.

Within the NAS, many of the lessons learned from the implementation of CPDLC and NEXCOM in the en route environment will transfer to the terminal environment. For example, the symbols used to indicate that an aircraft is data-link equipped, the methods available to select an aircraft for transmission, and the methods available to select or compose the message to be sent that prove successful in the en route environment, should also be appropriate for the TRACON displays. However, display clutter is generally more of a problem in terminal sectors than en route sectors, due to the increased density of the traffic. For this reason, the specific ways in which each additional piece of information (such as whether or not an aircraft is data-link equipped or whether or not a link had been established) to be displayed to the controller needs to be determined to be suitable for the TRACON displays into which they will be integrated (STARS or ACD). Other aspects of CPDLC that would be expected to be different from en route (and thus will need to be specifically evaluated in the terminal setting) include: the appropriate message set (so that the most frequently used messages are easily accessible); and the setting of the "time out" parameter (terminal communications are generally more time-critical than en route). The feasibility (cost/benefit) of integrating CPDLC with FAST tools (such as pFAST) should be explored so that information such as runway assignments could be data-linked to the cockpit at the controller's discretion.

NEXCOM is being developed to accommodate both voice and data link more efficiently than today's equipment. Like CPDLC, NEXCOM is also scheduled to be implemented in the en route environment before it is implemented in the terminal environment. According to the NAS Architecture, CPDLC Build 3 is scheduled to be implemented within the NEXCOM network in high density terminal sectors between 2007 – 2015. NEXCOM is in the early stages of development and the HMI for this system has not yet been defined.

One important operational requirement that NEXCOM may not be able to satisfy is the need to eliminate blocked and partially blocked (i.e., "stepped on") voice transmissions. As the amount of air traffic and radio frequency congestion increases, blocked and partially blocked transmissions present an increasing risk to aviation safety. When a pilot or controller is not able to access a frequency due to a "stuck mike," the most fundamental safety net - that provided by voice communications between pilots and controllers - is gone. Partially-blocked or "stepped-on" transmissions are far more common than microphones stuck in the transmit position; while these events are typically less dramatic than that of a stuck mike, they too, can contribute to controller and pilot workload and errors.

No studies have been conducted to examine the incidence of blocked and partially-blocked communications in today's ATC environment. However, in a high-fidelity en route simulation study designed to assess the level of communication delay that would be acceptable to controllers in the NEXCOM system, blocked communications (by pilots and controllers combined) were measured at 10 percent at the lowest communication delays (Sollenberger, McAnulty, and Kerns, 2002). Also, the number of step-ons is known to increase as the number of communications increases and with the amount of delay (between the onset of the speakers voice and the beginning of the transmission as heard by the listener) inherent in the system (Nadler, et al., 1990). As traffic continues to increase, the amount of frequency congestion and problems associated with blocked communications will continue to escalate.

In a 1998 study of communication errors reported to the Aviation Safety Reporting System (ASRS), blocked communications was identified as a factor that contributed to runway transgressions, altitude deviations, loss of standard separation, and pilots accepting a clearance intended for another aircraft. While similar-sounding call signs are the number one contributing factor to a pilot accepting a clearance intended for another aircraft and other critical communication errors, the risk of a blocked or partially-blocked transmission can compound the problem. When the "wrong" aircraft accepts a clearance, the pilot's readback can alert the controller (and other pilot) of the misunderstanding – as long as the readback contains a call sign and is not blocked. If two pilots respond simultaneously – as one would expect in the situation where two pilots think the clearance is for them – at least one readback is likely to be blocked.

While NEXCOM is proposed to incorporate anti-blocking capability, the precise technology that will be used to afford this capability has not yet been defined. There are also various ways to prevent blocked transmissions, and the operational suitability to the proposed method will need to be explored. For example, there are important operational differences between systems that:

- Puts a conflicting incoming transmission into a buffer, vs one that prevents step-ons by allowing the party (pilot or controller) who is trying to transmit hear the transmission that they would have stepped on. Furthermore, the implementation schedule does not project this (undefined and untested) capability to be available at airports before the year 2015.

- Prevents step-ons by displaying a "busy signal."

- Allows the party (pilot or controller) who is attempting to transmit to hear the transmission that the pilot or controller would have stepped on.

One possibility is that NEXCOM will use the frequency occupied indicator or "busy signal" for the pilot and have continuous controller override. There are two potential problems with this. First, it does not afford the possibility of the pilot or the controller to hear the transmission that they would have stepped on. This information is much more valuable for the pilot than a "busy signal." First, listening to the content of the message can provide a cue as to when the frequency will be free, in addition to possibly increasing the pilot's situational awareness for the traffic situation. Second, an automatic override for the controller with a "busy signal" for the pilot deprives the controller and pilot of the ability to choose whether or not to block the incoming transmission. Since mostly all pilot-initiated messages, such as initial check-ins, are not time-critical, it is unlikely that a pilot would choose to block another transmission. And, while it is likely that controllers would choose to override incoming transmissions, there may be an operational requirement to give them the information needed to be able to choose to do so (such as a system that allows them to hear the transmission that they are about to step-on).

The implementation schedule does not project this (undefined and untested) capability to be available at airports before the year 2015. Projections in the rate of air traffic and the concomitant increase in frequency congestion make such a schedule problematic in addressing the future needs of the NAS, and incompatible with a commitment to reducing surface incidents.[2]

4.5 ENHANCED SURVEILLANCE

4.5.1 System Description

Enhanced surveillance technologies, such as those based on automatic dependent surveillance-broadcast (ADS-B) have the potential to improve capacity, efficiency, and safety along a number of dimensions. It will avail the controller of more accurate position information (for equipped aircraft) and will support enhanced cockpit displays of traffic information (CDTI).

4.5.2 Integration Issues to be Considered

The availability of ADS-B information has several integration issues associated with it. First, decisions will need to be made as to how and when to display ADS-B position data to controllers. For aircraft equipped with ADS-B, their broadcast position could be slightly different than the position reported by radar, due to the different update rates. These positions will need to be reconciled so that only one position is displayed for a single aircraft. Raytheon has successfully prototyped fusion of ADS-B and reports from multiple radar sites in STARS, however, how this information will be presented and associated procedural issues are still being explored (Bacon, Glaiel, Stamm, Jagodnik, Hasan, 2002).

Since not all aircraft will be ADS-B equipped, controllers will also need to know which aircraft are ADS-B equipped (or whether the aircraft position is being reported by ADS-B or radar returns); this will add another bit of information that will need to be displayed. Extensive human factors work is currently underway to address these issues within the ADS-B program.

ADS-B is also assumed to be an enabling technology for a progression toward "free flight." This progression consists of increasing degrees of pilots accepting responsibility for the safe separation of their aircraft from other aircraft. The principles of operation for use of airborne separation assurance systems defined by FAA/Eurocontrol (2001) define four levels of separation assurance applications. The first level of systems would be used solely to enhance the flight crew's awareness of the traffic situation. The second level involved the pilot achieving and maintaining spacing with designated aircraft. With the third level, the flight crew maintains separation from specified aircraft under limited conditions. With the fourth level, pilots maintain separation from all surrounding aircraft

---

[2] Meanwhile, anti-blocking technology has been in use at an ATC facility in the U.K. for several years. It has also been installed in aircraft by Austrian Air and Britania. Installation of such technology has long been advocated by the Allied Pilots Association and the Air Line Pilots Association.

Under the Program of Safe Flight 21, cargo aircraft used ADS-B technology to identify other aircraft on their traffic display in the cockpit, and follow that aircraft at a specified distance. By "self-spacing," pilots assume some of the responsibility for safe separation that has traditionally been assumed by the controllers. The human factors issues associated with this shared separation responsibility are extensive and beyond the scope of this document. Critical issues include a determination of the pilot's willingness and ability to perform such a task and the controller's ability to maintain adequate awareness of these aircraft so that control can be resumed at any time. These issues will first be addressed in contained areas of low traffic density. Extensive testing will be required to determine whether such procedures are viable, or of any operational benefit, in the terminal environment. If so, interoperability with all systems to be integrated with ADS-B, such as pFAST and data link, will need to be closely examined. It would be useful if conflict detection and resolution tools used by pilots and controllers could accept ADS-B information as well as radar information and had the ability to differentiate between them.

Finally, controllers will need to fully understand the limitations of the ADS-B based information presented and be informed of any known degradations or failures. For example, what is the likelihood that a target will be "dropped" or misidentified, or that a false target will be displayed? Controllers and pilots will need to be informed of any failure or degradation in the accuracy of ADS-B. Back-up procedures, to be used in the event of a failure, will need to be developed for pilots and controllers to use when pilots are maneuvering based on ADS-B information.

4.6 ENHANCED TRAFFIC MANAGEMENT SYSTEM (ETMS) UPGRADE

Enhanced Traffic Management System (ETMS) was originally designed for and used by TMUs. However, the NAS Architecture 4.0 and the 1999 Capital Investment Plan describe enhancements to ETMS to permit the display of ETMS information on the controller's display. Such displayed information would require the integration of ETMS into STARS. This section explores the issues associated with the integration of ETMS into STARS and ACD.

4.6.1 System Description and Integration Issues to be Considered

The traffic situation display (TSD) graphically depicts current aircraft position superimposed on maps of geographical boundaries and NAS facilities. (However, all of the boundaries, airways, fixes, and other geographical information is already available on STARS.) ETMS also displays terminal weather. This weather information is supplied to the ETMS by the Environmental Research Laboratories, and would not be as useful to terminal controllers as the information from the Integrated Terminal Weather System. While the simultaneous display of traffic and weather is extremely useful for both tactical and strategic decision-making, the integration of the weather information from the ITWS onto STARS satisfies this requirement. ETMS simultaneous display of traffic and weather information would be useful to those facilities that would not otherwise have access to such weather information.

The ETMS system will also include (with the implementation of the Collaborative Routing Coordination Tools - CRCT), the following functionality:

- Display ground delay program status reports and ground delay histories.

- Project traffic demands for a specific airport, sector or fix and generates an alert when the projected demand exceeds the alert threshold.

- Generate and display traffic problem resolution strategies for individual or groups of aircraft.

- Assess the impact of reroutes on sector volume, aircraft spacing, and traffic density.

Some of this information will be available to TMUs with CTAS (Center/TRACON Automation System). For example, the Traffic Load Graph Display shows the number of aircraft predicted to: enter TRACON airspace in a fixed time interval, cross a runway threshold, final approach fix or meter fix. However, this information is displayed as a graph with time on the x-axis and number of aircraft on the y-axis. Also, this information is not sector-specific. The CTAS Planview Graphical User Interface does provide a spatial display of individual aircraft track information, that is, individual aircraft on a predefined airspace radar map. The information available for display with this interface consists of:

- Aircraft symbols
- Flight datablocks
- Waypoints
- Range rings
- UTC clock
- Scratch pad

The display of the aircraft symbols is automatic; display of the other items is optional. While sector boundaries are not currently available, they could be added in.

The Timeline Graphical User Interface of pFAST includes the capability to display:

- Timeline displays of traffic approaching specific reference points.

- Load graph displays of traffic scheduled to cross-specified reference points in a given period of time.

- The scheduled delay for individual aircraft.

- The impact of changes in airport configurations, acceptance rates, or other scheduling constraints.

The usefulness of this information provided by CTAS is limited by several factors. First, in the planned implementation, only the TMUs, not individual controllers, will have access to it. Second, not all facilities will have these CTAS tools implemented. Finally, some of the information is presented only in a timeline format. That is, it cannot present the information in a situation display mode. This limitation is not inherent in the ETMS. ETMS shows the actual traffic and "look ahead" views as flows of traffic, showing position and direction (as opposed to a graph showing the numbers of aircraft as a function of time). The flow and complexity of traffic can be displayed for all traffic or specified groups of aircraft (e.g., east bound, by destination or waypoint). Displaying aircraft by type (e.g., jets in one color, props in another color) could help controllers plan arrival streams more efficiently. This ability to look ahead and project traffic demands would be useful to individual TRACON controllers and to a tower supervisor or controller-in-charge (CIC) for planning purposes. In the tower, the ability to look ahead and view the traffic 60 miles from the airport could be used to make decisions about required staffing (e.g., if one or two local positions will be required), to project the effects of gate holds (i.e., determine whether incoming aircraft have a place to park), to time changes in runway configurations, and manage traffic more efficiently.

Making ETMS information with the anticipated enhancement of the CRCT tools available to TRACON controllers would enable the controllers at each sector to anticipate traffic demands with a sector-specific graphical display of the aircraft coming into the sector at timed intervals, and display the predicted effects of reroutes (e.g., around weather). While these functions are often performed by TMUs, it is important to note that not all TRACONs have TMUs. Also, TMUs do not have the capability to convey this information to individual controllers in a graphical display of traffic. Similarly, not all facilities will have pFAST (i.e., the modified data block) to assist TRACON controllers in determining the most efficient arrival streams. Furthermore, the capabilities of ETMS/CRCT to: alert the controller to projected demands that exceed given thresholds, assess the impact of reroutes, and generate conflict resolutions would be useful – assuming that these functions are easy to use and perform as advertised. Finally, it is anticipated that the ability to obtain a timed look-ahead at traffic demand will become increasingly important as the science of air traffic management in terminal areas becomes more strategic, with less reliance on sector boundaries, and more focused on the flexibility needed to entertain concepts such as "free flight."

It is important to note that while ETMS information can enhance situation awareness of controllers in the TRACON and tower, and has the potential for improving efficiency of terminal operations, it is not required for the controllers' primary task of providing safe separation. There is a tendency on the part of users to think that more information is better – at least until they try to use the system with the new information on it. A guiding principal of display design is that only information that is immediately useful is displayed and it is displayed in a format that is immediately useful (i.e., no mental gymnastics required). Other information, such as that that would be required with a change in circumstances (e.g., aircraft declares an emergency), or would be useful as a tool, should be available to be called up as needed, but not be constantly present in the primary field of view (i.e., the center of the display of traffic). Currently, display clutter can be a problem even when only necessary information is presented. Display of unnecessary information must be avoided at all costs as it can distract from, or obscure, critical information.

ETMS with CRCT does contain information and tools that could be useful to TRACON controllers. However, since ETMS information is not required for the tactical, time-critical decisions required of these controllers, the complex integration required to implement

ETMS/CRCT functionality into STARS is not justified. In fact, such integration could interfere with other functionality by unduly complicating the STARS user interface. However, there are several alternatives to integrating ETMS into STARS, which would still make the information accessible to TRACON controllers. For example, an ETMS station could be set up between controller workstations. Another alternative is one that is currently being used at Oakland Center. While most facilities have an ETMS display only at the TMUs and/or supervisor's station, Oakland Center has an ETMS station at each bay of (6) controllers. This allows the controllers to collectively see the aircraft in their (and their neighbors') airspace as well as (have their D-side controller) query the system for specific information, as needed. This displays customized ETMS data for a small group of controllers by projecting it on a wall mounted (roughly 4' x 8') screen and maintains the capability of querying the system. Such alternatives, that would avail TRACON controllers and tower supervisors of the ETMS/CRCT tools, without integrating the system into the already complex STARS interface, should be seriously considered.

This page intentionally left blank

5. THE NEW FACE OF HUMAN-SYSTEM INTEGRATION (HSI)

In the integration of decision support tools or enhanced forms of information (such as higher quality weather information), controller "acceptance" is often the singular goal of the program. While it is appropriate for this to be a critical factor in determining how systems are integrated into the NAS, it should not be the driving force in system development or in system integration. The primary question, to be answered before any new system is considered for implementation, is: What are the user's information requirements? This includes: Why is the information needed? How will the information be used? The answers to these questions will help to determine what information is presented (vs. available or not available) and how it is presented. After these questions are answered, assessment is required, not to determine what a group of user representatives consider acceptable, but to determine how different display alternatives will affect performance.

The over-reliance on user acceptance can be historically traced to the early days of STARS. The corporate culture within the FAA at that time was that STARS was, above all, to be a Commercially Off-The-Shelf (COTS) acquisition. This precluded any tailoring to user requirements, despite the fact that the characteristics of the NAS airspace in which the system was to be used was operationally very different from the airspace in which the existing COTS systems were in use. In early discussions of operational (including human factors) specifications, requirements that could not be met by the existing systems, or that would require the manufacturer to demonstrate that the new system was at least as effective (e.g., induce no additional workload or not increase the potential for errors) as the existing system, were categorically and intentionally excluded. Also excluded was structured input from a wide segment of the user population. This led to conflict between the FAA and the National Air Traffic Controller's Association (NATCA) in which attention to human factors was seen as the central issue. Congress and the Inspector General of the Department of Transportation intervened and a human factors assessment of STARS was mandated to be completed within 90 days. Considerable effort was expended by many talented individuals from several organizations to ensure that the evaluation was the best that it could be in the time allotted. However, as is always the case when schedule is the driving factor, the effort was severely constrained. What could have, and should have, been an extensive human factors evaluation of STARS was limited, by time and funding to an assessment of a finite number of aspects of the system. Nonetheless, this set the stage toward a more formal identification and resolution of human factors issues. Human factors was becoming recognized as a tool for identifying and resolving issues of both usability and user acceptance.

While usability and user acceptance are clearly related, the two are not interchangeable. In addition to identifying and resolving issues of usability and user acceptance, the tools of the science of human factors can, and should, also be used to help determine whether an issue is one of usability or acceptance. This helps to ensure that issues are put in proper perspective and interpreted correctly. When issues are identified by user groups as ones that affect operational acceptability without an objective assessment of the effect on measures of usability (such as error rates and response times), when preference takes precedent over science, every one -- from the potential user of the system to the taxpayer -- loses. As human factors came to the forefront of

STARS, the role of the human factors specialists evolved. Traditionally, human factors specialists focused on ensuring:

- That the system was designed to minimize human error and maximize efficiency.

- That the system allowed for errors to be detected and corrected quickly and easily.

- Proper allocation of function between the operator and the system.

- That the effects of system use on training and staffing requirements were adequately addressed.

While the need for these activities has not diminished, an added role was assigned to the human factors specialists on an evaluation team, that of "consensus building." The definition of programmatic success had shifted toward getting an identified group of users to agree on what was acceptable, what needed to be changed, and what alternatives would be accepted by the group. With this, came a shift from rigorous testing of specific alternatives toward demonstrations of capabilities. Instead of helping operations specialists with early identification of human factors issues and deciding how these issues should be examined and resolved, human factors specialists were now charged with helping operations specialists reach consensus on what would be deemed acceptable. From a programmatic standpoint, this is problematic for several reasons.

It is a well-known principle of human factors that people are not the best predictors of how display options will affect their performance. With many aspects of computer-based systems, it has often been the case that people prefer one set of options when they have unknowingly performed better with another set of options (see Andre and Wickens, 1995 for an interesting review). Such cases of this preference-performance disassociation have not been as well documented in the realm of aviation. However, there are many anecdotal examples in the initial stages of system development (e.g., cockpit displays of traffic information) in which users stated that they preferred having more information than they would later prove to be able to use. The types of preference-based vs. performance-based decisions that are most vulnerable with respect to accuracy, appear to be those regarding how much information to present and how the information should be coded. The use of color is particularly problematic as people consistently prefer the use of more color-coding than can be shown to enhance performance; in fact, some color displays have been preferred by users even when they (unknowingly) degraded performance.

Factors that affect preferences are different from factors that affect performance: they are more variable, more likely to change from person to person, and are more subject to change with operational experience than factors that affect performance. Operational experience refers not only to the level of experience that the controller has in general, but also to the level of experience with specific systems. For example, excellent work was done by MITRE to determine the best way to present controllers with the Resolution Advisories (RA) presented to pilots from the TCAS (Hoffman, Kaye, Sacher, and Carlson, 1995). At the time of the study, it was assumed (and correctly so) that controllers wanted to know what the RA was that was presented to the pilot so that they would know when and why an aircraft would purposely deviate from an assigned altitude. However, this line of research could be said to have started with the wrong question. The primary question should have been "Does the controller need to see the RA and if so, how will the information be used?" Instead, the operational question addressed was how to best present the information to the controller. When TCAS was first implemented, there were a variety of misconceptions surrounding, and hence a general distrust of, the TCAS system by controllers. In time, controllers learned more about how TCAS would operate; with this increased understanding came a decreased apprehension. A variety of factors came together to change controllers' attitude toward TCAS. However, interviews with controllers conducted in the context of this study indicated a shift in controller preference: all of the controllers' interviewed said that they would prefer that the RA not be presented in the data block. Note: this does not shed light on whether or not the presentation of the RA would positively or negatively affect controller performance; that study has not yet been done. Nor does it address the question of what, if anything, controllers need to know (from the ground system as opposed to from the pilots) when pilots get an RA. Rather, it suggests a shift in controller preference as a result of operational experience with TCAS (and all of the factors that went along with this - such as the resolution of controller liability concerns, enhancements to the TCAS software that reduced the number of "nuisance" RAs, etc.).

In addition to the preference-performance disassociation, there is also another disconnect that occurs with ATC systems, that is, the definition of success in engineering, programmatic, and operational terms. From an engineering standpoint, a system is a success if it performs as it was intended to perform (e.g., balances the traffic across fixes or runways). From a programmatic perspective, success could be defined as the system being operational (i.e., turned on) at a few sectors in a single facility at a certain time of day. While this is a very limited operational experience, it could be deemed a programmatic success if it is achieved within a scheduled deadline. From an operational perspective, however, the system will only be deemed a success if it satisfies an operational need in a manner that is compatible with the controllers tasks. If a decision support tool is not perceived as being highly accurate, reliable and easy to use, it will not be used by controllers as the developers intended it to be used. Controllers will have a much lower tolerance (in terms of accuracy, reliability, workload required to use the system, etc.) for systems that do not benefit them directly.

A critical question to be asked is: "Is this something that controllers need or can reasonably be expected to want to use?" This is determined by how the use of the system will affect the controllers' tasks and responsibilities (liabilities). A key issue here is the proper allocation of function. Any automated function that is perceived as taking control or flexibility away from the controller will be met with resistance. Similarly, the tool must be perceived as being useful to the controller and worth the time and energy required to use it. "Altruistic" systems that do not directly benefit the controller using the system will also be met with resistance. This experience has been borne out with several systems, including the Departure Sequencing Program (DSP). DSP provided tower controllers with a constrained takeoff window for departing aircraft; compliance with these times would help to reduce congestion at the fixes (outside of tower airspace). DSP was deemed an engineering success from the start, as compliance with the take-off windows did result in balanced traffic over the departure fixes. However, it was not an operational success until several changes were made. For example, the initial version of the program required controllers to enter information into the system whenever departure windows were not able to be met; the system in use today updates automatically. This tower-based system (used only when traffic demand exceeds capacity) now also displays departure information at the D-side position in some TRACONS to aid in planning.

Similarly, pFAST gave controllers an assigned runway and sequence to the runway to help balance the traffic into the airport onto the different runways. As with DSP, there was no return on investment for the controller charged with the necessary care and feeding of the computer.[3] Both systems provided additional workload, constrained the controller's options in terms of decision-making and, in the case of DSP, provided no useful information to the user; these are the results of a misplaced allocation of function between the system and the user.

It has become clear that *how* the system is introduced to controllers is as important as *what* is introduced. One of the lessons learned in the implementation of pFAST was that the controller user group needs consistent support in identifying issues and "translating" them into the design or parameter changes to be implemented by the engineers. Human factors support is also useful in working with engineers to identify options that satisfy controller requirements. The tools of the science of human factors have long been used to help separate fact from opinion. They can be used to identify characteristics of a display or data entry procedures that can induce errors, independent of user opinion or preference; design studies to determine whether changes to a display will affect performance; determine whether differences found in a study are more likely to be due to chance or one of the variables tested; and ensure proper interpretation of study results.

One of the ways of enhancing the probability of successful implementation of a new system is to have a consistent team of human factors specialists and engineers to support the project from the introduction of a system to operational testing and evaluation. This team would work with the NATCA national team formed to address integration issues and would conduct the following activities:

1. Assess individual facility requirements (i.e., what functions will be useful at this facility?)

2. Discuss capabilities and limitations of system with local facility representatives (NATCA, TMU, facility managers, training personnel). This is a vital step toward helping to manage user expectations and ensure that they are realistic.

3. Identify specific characteristics of the facility's operation that will need to be accommodated. (If the system has been implemented at another facility, care should be taken to address aspects of the operation that differ between the two facilities.)

4. Support the initial testing phase of the implementation.

---

[3] An interesting exception would occur with "up/down" facilities in which controllers rotate through TRACON and tower positions. For example, if pFAST had undergone a more extensive trial period at ATL, TRACON controllers would have been bale to experience the benefits to tower operations first hand as thy worked the positions in the tower.

5. Develop a compendium of controller issues/concerns. It is often the case that the issues identified vary with the individuals involved. As the make-up of the user groups change over time, there can be a cost – in terms of time and other resources – of revisiting decisions that were made in the past. Detailed accounts of the decisions made (and the rationales for those decisions) regarding the capabilities to be implemented and how they will be implemented need to be documented and tracked. Human factors specialists can be very useful in this process, including:

- helping controllers identify and verbalize operational issues with the new system,
- helping to differentiate between operational requirements and "nice to haves,"
- translating the operational requirements into engineering options, and
- ensuring that the site adaptation has been adequately addressed.

This document has identified specific issues that need to be considered in the implementation of enhancements to the STARS and ACD legacy systems. This does not detract from the need to give each, individual facility human factors support in their system integration efforts. It is well known that the specific adaptation of a system (such as pFAST or AMASS) to support the unique characteristics of a facility's operation is as costly as it is critical to the effective operation of the system at that facility. However, this is not the only way in which a facility will need human factors support for successful implementation. Each facility is different in terms of their operations, local procedures, and local "culture." Furthermore, not all facilities are scheduled to get many of the sub-systems discussed. Each different combination of subsystems and facility characteristics creates a unique set of integration issues. For this reason, there will be specific integration decisions that will need to be made on a facility-by-facility basis. Furthermore, any changes in airspace or airport configuration (such as the addition of a runway) that affects the operation of a sub-system will require refinement of that sub-system and testing to ensure that the operational requirements continue to be met. The importance of providing such support to each facility cannot be overstated.

This page intentionally left blank

6. SUMMARY OF EXISTING CHALLENGES AND SUGGESTIONS FOR FURTHER RESEARCH

This document identifies existing human factor challenges to the realization of system benefits with the integration of enhancements into the TRACON environment and presents suggestions for research that will help define operational requirements for system integration. This section presents a summary of those issues.

Existing Human Factor Challenges

- In order to realize the benefits projected by any subsystem, it must be used consistently in the manner intended by the designers. This requires that the user consider the system to be acceptable, trust-worthy, and provide a useful function. In addition to satisfying well-known human factors requirements for system performance, the system must also be able to operate effectively in all operational conditions, such as various wind and weather conditions, different runway configurations, with changing spacing requirements, etc. This challenge was highlighted by the operational experience with pFAST that should continue to be refined.

- All decision support tools will need to ensure that the decision authority remains with the controller and is able to be used effectively within a range of operational conditions. The benefits and limitations on the interactions between the users of the system and others with whom they interact will need to be determined. This includes the coordination performed between TRACON controllers and the tower, ARTCCs, other TRACONs, coordination among different TRACON positions, the position relief briefing (within a position), and controllers and pilots. Specific interactions include: the use of pFAST on tower operations and on the incidence of runway incursions, and the use of CRDA on tower controller-pilot voice communications and other aspects of tower operations.

- Data recording capabilities will need to be able to capture display and control variables (such as preference settings) to support the investigation of specific variables on operational errors. For example, if aircraft can be presented in different colors and/or at different brightness levels, such variables will need to be able to be recorded to support a retrospective analysis of critical events.

- Tools that increase traffic efficiency often have a concomitant increase in traffic complexity. Therefore, the potential effects of the use of the tool, and any changes in the presentation of information to the controller, on operational errors will need to be anticipated and monitored.

Suggestions for Further Research

- Information Coding. Research is needed to evaluate different techniques to help the controller highlight some aircraft (e.g., to identify "their" data blocks) without increasing the probability that the controller would fail to detect a potential conflict between highlighted and non-highlighted aircraft. The degrees of difference in intensity (brightness) and in color that could affect the controller's ability to detect potential conflicts needs to be empirically determined.

- Weather. Research, such as a task analysis, is needed to determine the controllers' operational requirements for weather information. This would provide the basis for selecting the information to be displayed (as opposed to available) to the controller. This should be differentiated from the Traffic Management Unit's requirements for weather information.

- Communication. Continued research is needed to determine the viability of the use of data link communications in the terminal environment. That is, what types of communications can be supported by data link (as opposed to supported by voice). Work is also needed to project the operational requirements of pilot-controller communications in the terminal environment and determine whether planned enhancements will be able to fulfill these needs. This includes the determination of the incidence of blocked communications in today's environments, the projection of this incidence with increased traffic density, and the projected effects of specific implementation options, such as frequency override capabilities and the method used to prevent blocked-transmissions.

- Enhanced Traffic Management System (ETMS). While ETMS information is not required for the tactical, time-critical decisions required of TRACON controllers, ETMS with CRCT does contain information and tools that would be useful in helping controllers anticipate traffic demands and predict the effects of reroutes (e.g., around weather). An ETMS station between controller workstations, or at each bay of controllers, would allow the controllers to collectively see the aircraft in their (and their neighbors') airspace as well as have their D-side controller query the system for specific information, as needed. The feasibility, costs, and benefits of making an Enhanced Traffic Management System (ETMS) Display accessible to TRACON controllers should be investigated.

- Free-Flight. Research is needed along several fronts in order to support the development toward "free flight."

  - Whether and how the use of ADS-B information will affect controller tasks and procedures must be carefully examined.

  - The extent of pilots' willingness and ability to perform various degrees of self-separation and assume responsibility for separation assurance needs to be determined.

- The degree to which current flight deck capabilities can support traffic awareness and self-spacing in the terminal environment.

- The controllers' ability to maintain sufficient situation awareness to identify and intervene in potential conflicts under conditions of degrees of pilot self-separation needs to continue to be investigated.

- As advanced cockpit systems are developed and deployed to support various degrees of "free flight," research must be conducted to determine what information, if any, should be downlinked from these systems to controllers.

This page intentionally left blank

APPENDIX

Ambient Illuminance Measurements in foot candles (fc) for Three TRACON Facilities – August 2002

	Dallas TRACON		Atlanta TRACON		El Paso TRACON	
	At Seat	At Screen	At Seat	At Screen	At Seat	At Screen
Mean	.09	.06	2.71	1.6	11.15	3.36
St. Dev.	.02	.01	.59	.15	3.84	NA
Minimum	.02	.04	1.7	1.2	5.15	2.77
Maximum	.15	.07	3.8	1.6	14.9	3.94
Number of Measurements Taken	30	11	17	6	4	2

At Seat – measurement taken at or near at controllers head.
At Screen – measurement taken at the screen

Method:

Illumiance measurements were taken using a hand held illuminance meter (EXTECH Model 401036 Light Meter). The calibration of the meter was checked against a second portable illuminance meter that had recently been calibrated against NIST traceable standards. Measurements taken with the EXTECH unit were found to be similar, but generally higher than comparable measurements taken with the calibrated unit. At the lowest light levels, the EXTECH unit yielded readings 50 percent higher than the calibrated unit. At higher light levels (approx 25 fc) the EXTECH yielded readings only 15 percent higher. These differences are believed to be due primarily to differences in the acceptance angle of the sensors in the respective units. The data reported here are best used for relative comparisons between facilities.

An attempt was made at each facility to gather measurements at both the head position of the controllers and at the level of the ACD or STARS display. It was not possible to get 'At Screen' measurements at active workstations, so the sample size is smaller for these columns.

This page intentionally left blank

GLOSSARY

ACD	ARTS Color Display
ADS-B	Automatic Dependent Surveillance-Broadcast
aFAST	active Final Approach Spacing Tool
ARTCC	Air Route Traffic Control Center
ARTS	Common automated Radar Terminal System
ASRS	Aviation Safety Reporting System
ATC	Air Traffic Control
CA	Conflict Alert
CASA	Controller Automation Spacing Aid
CDTI	Cockpit Displays of Traffic Information
CIC	Controller-in-Charge
CPDLC	Controller-Pilot Data Link Communication
CRCT	Collaborative Routing Coordination Tools
CRDA	Converging Runway Display Aid
CTAS	Center TRACON Automation System
DA	Descent Advisor
DSP	Departure Sequencing Program
DST	Decision Support Tool
ETMS	Enhanced Traffic Management System
HMI	Human-Machine Interface
IFTA	Integrated Turbulence Forecast Algorithm
ITWS	Integrated Terminal Weather System

MSAW	Minimum Safe Altitude Warning
NAS	National Airspace System
NATCA	National Air Traffic Controllers Association
NEXCOM	Next Generation Air-Ground Communication System
NTZ	No Transgression Zone
P³Is	Pre-planned Product Improvements
PASS	Professional Airway System Specialists
pFAST	passive Final Approach Spacing Tool
PRM	Precision Runway Monitor
RA	Resolution Advisories
STARS	Standard Terminal Automation Replacement System
TCAS	Traffic Alert and Collision Avoidance System
TMU	Traffic Management Unit
TRACON	Terminal Radar Approach Control
TSD	Traffic Situation Display
URET	User Request Evaluation Tool
VDL	VHF Digital Link

REFERENCES

Allendoerfer, K., Bacon, E., Bohne, A., and Freitag, S., *Advanced Weather Products on Terminal Automation Systems*. Proceedings of the 46th Annual Air Traffic Control Association Conference, Fall 2001, 88-92.

Andre, A., and Wickens, C. When Users Want What's NOT Best for Them. *Ergonomics in Design*, October 1995, 10-13.

Bacon, E., Glaiel, F., Stamm, R., Jagodnik, A., Hasan, R., *Automatic Dependent Surveillance – Broadband (ADS-B) Prototyping on STARS*. Proceedings of the 47th Annual Air Traffic Control Association Conference, November 3-7, 2002, 102-106.

Beal, W., Reid, w., and Schlimper, *The Operational Performance of the Safety Functions Conflict Alert and MSAW*. Proceedings of the 45th Annual Air Traffic Control Association Conference, Fall 2000, 85-89.

Burnett, K., Beasley, P., and Mundra, A.,*Converging Runway Display Aid as a Means to Increase Airport Capacity*, 2000. Proceedings of the 45th Annual Air Traffic Control Association Conference, Fall 2000, 232-236.

Cardosi, K. *An Analysis of En Route Controller-Pilot Voice Communications*. 1993, DOT/FAA/RD-93-11.

Cardosi, K., Brett, B., and Han, S. *An Analysis of TRACON (Terminal Radar Approach Control) Controller-Pilot Voice Communications*. 1996, DOT/FAA/AR-96/66.

Cardosi, K., and Hannon, D., *Recommendations for the Use of Color in ATC Displays*. 1999, DOT/FAA/AR-99/52.

CTAS Final Approach Spacing Tool, Passive FAST Reference Manual. Document #ARC-FAST-UM-98-001.2 Release 5.5.0 Dated 4/14/2000.

Davis, T., Isaacson, D., Robinson, J., den Braven, W., Lee, K., and Sanford, B., *Operational Test Results of Passive Final Approach Spacing Tool*. 1997. Proceedings of the IFAC 8th Symposium on Transportation Systems, Chania, Greece, June 1997.

Dimeo, K., Sollenberger, R., Kopardekar, P., Lozito, S., Mackintosh, M., Cardosi, K., and McCloy, T. Air-Ground Integration Experiment. January 2002, DOT/FAA/CT-TN02/06.

Enhanced Traffic Management System (ETMS) Functional Description Version 5.0 Report No. Volpe Center DTS56-TMS-002. June 30, 1995.

Federal Aviation Administration, ACB-220. *User Interface Designs for Advanced Weather Products on Terminal Air Traffic Control Displays*. Draft Report ACB2202002-02, September 2002.

Federal Aviation Administration/Eurocontrol. Principles of Operation for Use of Airborne Separation Assurance Systems. Version 7.1. June 19, 2001.

Free Flight Program Office. Free Flight Phase 2 Research Program Plan (November 2001).

Hoffman, R., Kaye, R., Sacher, B., and Carlson, L. TCAS II Resolution Advisories Downlink Evaluation Report. August 1995. MITRE Report No. MTR-95 W 0000080.

Jones, H., *Activities of the Federal Aviation Administration's Aviation Weather Research Program.* Proceedings of the 46th Annual Air Traffic Control Association Conference, Fall 2001, 123-126

Kabaservice, T., *A Proposed Capability Demonstration for Next-Generation Digital Air-Ground Voice Communications.* Proceedings of the 43rd Annual Air Traffic Control Association Conference, Fall 1998, 60-64.

Landis, M. Internal Communication, July 31, 2002.

Lee, K.K. and Sanford, B.D., *The Passive Final Approach Spacing Tool (pFAST) Human Factors Operational Assessment.* 2nd USA/Europe Air Traffic Management R&D Seminar, Orlando, FL, 1-4 December 1998.

Meyer, E., Post, J., Blucher, M., Fralick, D., *An Operational Assessment of the Passive Final Approach Spacing Tool at Dallas-Ft. Worth International Airport.* Proceedings of the 45th Annual Air Traffic Control Association Conference, Fall 2000, 23-29.

Nadler, E., DiSario, R., Mengert, P., and Sussman, E., *A Simulation Study of the Effects of Communication Delay on Air Traffic Control.* 1990, DOT/FAA/CT-90/6.

Perry, T., *In Search of the Future of Air Traffic Control*, IEEE Spectrum, August 1997.

Ryan, H., Kazunas, S., Paglione, M., Cale, M., *Issues in the Design and Use of a Common Aircraft Trajectory Modeler.* Proceedings of the 42nd Annual Air Traffic Control Association Conference, Fall 1997, 72-76.

Sollenberger, R., McAnulty, M., and Kerns, K. *The Effect of Voice Communications Latency in High Density, Communications-Intensive, Airspace*, Draft Final Report, September 2002.

Wickens, C., Mavor, A., Parasuraman, R., and McGee, J., *The Future of Air Traffic Control: Human Operators and Automation*, The National Academy Press, 1998.

www.ingramcontent.com/pod-product-compliance
Lightning Source LLC
Chambersburg PA
CBHW081908170526
45167CB00007B/3205